ARDUINO

COMPLETE BEGINNERS GUIDE FOR ARDUINO

EVERYTHING YOU NEED TO KNOW TO GET STARTED

MATTHEW MCKINNON

TABLE OF CONTENTS

INTRODUCTION...5

CHAPTER ONE WHAT IS ARDUINO ? 7

CHAPTER TWO ARDUINO BOARDS.....................................11

CHAPTER THREE ARDUINO TERMINOLOGIES.................19

CHAPTER FOUR ARDUINO IDE...23

CHAPTER FIVE UNDERSTANDING ARDUINO SYNTAX.31

CHAPTER SIX PROGRAMMING EXPRESSIONS 35

CHAPTER SEVEN ARDUINO PROJECT CREATION
PROCESS 41

CHAPTER EIGHT DOWNLOAD AND INSTALL THE
ARDUINO IDE...45

CHAPTER NINE HARDWARE OVERVIEW 55

CHAPTER TEN ARDUINO COMMANDS...............................65

CHAPTER ELEVEN ADVANCED PROGRAMMING
CONCEPTS 69

CHAPTER TWELVE SAMPLE CODES FOR STUDY 91

INTRODUCTION

Are you eager to open up to the nifty world of Arduino boards?

Do you have a cool idea in your mind and looking for something to realize your ideas? Have you just bought an Arduino board, and want to make sure you get a running start? Then this book may help you in your endeavor.

It covers the information you need to jump in feet first:

- Arduino terminologies and different types of Arduino boards

- Arduino IDE

- Arduino syntax and Programming expressions

- Hardware overview and Arduino commands

- Advanced programming concepts – Interrupts, Array, Arduino library

- Sample codes for study and more

If you are eager to join in on the adventure of designing cool projects by using Arduino Boards, this book will help you accomplish it.

Thanks again for downloading this book, I hope you enjoy it!

CHAPTER ONE

WHAT IS ARDUINO ?

Arduino is a type of open-source platform which may be used to develop electronic projects. Arduino platform is made up of two separate components – microcontroller and an IDE. Microcontroller is also known as programmable circuit board which constitutes the hardware of Arduino. IDE is a software which runs on your system and it lets you write and upload code to microcontroller. Arduino is known as a physical or embedded computing platform which implies that arduino helps to sense the physical world better compared to our current computers. Arduino helps in the development of interactive objects by using the inputs from switches and sensors to regulate physical outputs.

Arduino can be considered as a tiny computer which enables to process inputs and outputs between the gadget and external modules you attach to it. For instance, you may use Arduino to turn on light for a pre-defined time, may be 10 seconds after we press a button. You may use this concept to perform a variety of purposes like creating digital clock, temperature and humidity sensors etc. All the design files for

Arduino board have been made public. Any person can make clones of Arduino since it is an open-source hardware.

Online Resource:

Arduino Products:
https://www.arduino.cc/en/Main/Products

Arduino Board Specifications:
https://www.arduino.cc/en/Products.Compare

Arduino Uno:
https://www.arduino.cc/en/Main/ArduinoBoardUno

Advantages of Arduino

- Affordable – Arduino board is available at affordable rates compared to other microcontroller platforms.

- Cross-platform support – Arduino backs Windows, Linux and Macintosh OSX.

- Easy to follow programming environment – Programming environment is simple and easy to follow. Moreover, the platform is simple and flexible which assists in accomplishing complex projects.

- Extensible software and hardware – Arduino is completely open-source and helps to expand the capabilities by using languages like C++, AVR C. It enables programmers to make module versions, enhancing, extending and the like.

- Program it via USB cable – Most of the modern day computers support USB cable and hence this is a really good feature.

- Active community of users – You can also get support from like-minded users who are experts in this field

Atmel Microcontroller

Main unit of Arduino is an Atmel microcontroller unit which is accountable for executing the commands you specify. Arduino Uno uses an AVR ATMega microcontroller whereas the Arduino Due uses an ARM Cortex microcontroller. A 16 Mhz ceramic resonator is attached to the clock pins which functions as the reference to execute your program commands. User can use reset button to restart program execution.

CHAPTER TWO

ARDUINO BOARDS

There are a wide variety of Arduino boards currently available in market and new boards are released with improved aspects regularly. Hence, let us concentrate on some of the leading official Arduino boards.

Arduino UNO

Arduino UNO is the widely used microcontroller board and is based on the ATmega328P. Arduino UNO encompasses 14 digital I/O pins, 6 analog inputs, 32 Kb flash memory and clock speed of 16 Mhz. Operating voltage is 5V whereas recommended input voltage is 7-12 V.

Fig 2.1 Arduino UNO

Arduino Leonardo

Arduino Leonardo uses ATmega32u4 and includes 20 digital I/O pins, 32 Kb flash memory and clock speed of 16 Mhz. Operating voltage is 5V whereas recommended input voltage is 7-12 V.

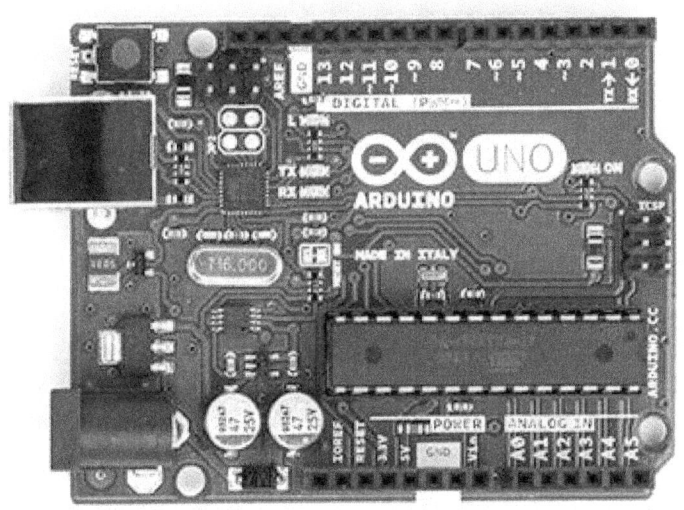

Fig 2.2 Arduino Leonardo

Arduino Mega 2560

Arduino Mega 2560 uses ATmega 2560 as the main microcontroller unit. Arduino Mega 2560 is conceived for complex projects like rapid prototyping. It includes 54 digital I/O pins, 16 analog input pins, flash memory of 256 kb and clock speed of 16 Mhz. It helps you to interface with multiple devices as it has 54 GPIO pins.

Fig 2.3 Arduino Mega 2560

Arduino Due

Arduino Due employs Atmel SAM3X8E ARM Cortex-M3 CPU as microcontroller. It encompasses 54 digital I/O pins, 12 analog inputs, 2 analog outputs and flash memory of 512 kb. Operating voltage is 3.3 V whereas recommended input voltage is 7-12 V.

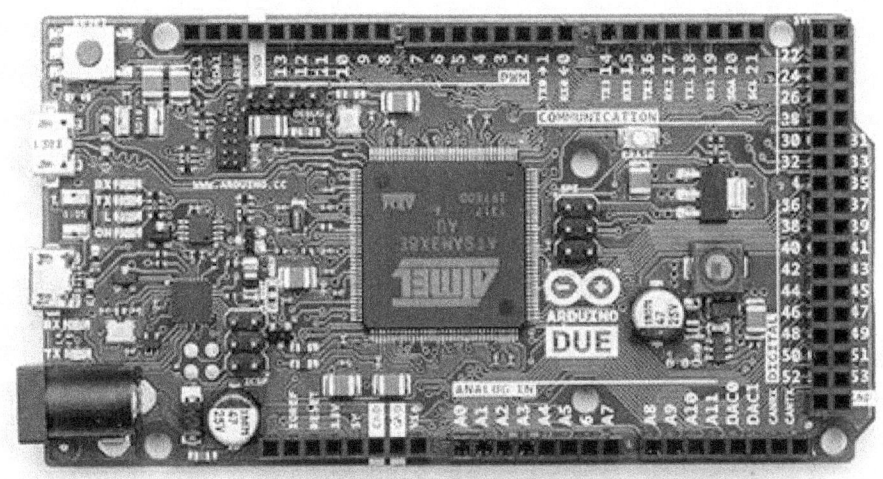

Fig 2.4 Arduino Due

Arduino Nano

Arduino Nano employs ATmega328 as the microcontroller unit. As the name implies, its small form factor makes it the ideal option for more finished projects.

Fig 2.5 Arduino Nano

Arduino Mega ADK

Arduino Mega ADK uses ATmega 2560 as the main microcontroller unit. It includes 54 digital I/O pins, 16 analog input pins, flash memory of 256 kb and clock speed of 16 Mhz. It helps you to interface with multiple devices as it has 54 GPIO pins.

Fig 2.6 Arduino Mega ADK

CHAPTER THREE

ARDUINO TERMINOLOGIES

Having good understanding of the terminologies and jargons is inevitable to get the most out of the book.

Breadboard – Breadboards are extremely useful for prototyping which is a reusable tool for developing circuits. It enables you to join circuits without permanently soldering them which serves as a stable surface for linking together modules.

Compiler – Compiler translates code that you have written into a form that Arduino microcontroller can perceive.

Device driver – Device driver is responsible for the communication between computer and gadget that you have linked to the system.

EEPROM – It stands for electrically erasable programmable read-only memory. Electric current has to be used to erase the chip content.

External interrupt – An external interrupt implies that some event has occurred outside the processor which demands attention.

Flash Memory – It is a type of memory which helps to retain data even if power is provided or not.

Digital Input/output – Digital pins can have either high or low value, where low means 0 or false and high means 1 or true. The number of digital pins can vary between different Arduino boards.

Analog Input/output – Analog signal means continuous electrical signal. Input for analog i/p pin is analog signal.

Processor – Computer instructions are fed into the processor which is then executed.

PWM pins – In PWM pin, data is transmitted by means of pulse – width modulation.

Serial communication – Serial communication occurs when two systems transmit digital pulses between them at a pre-defined rate.

Sketch – Sketch is analogous to code that we refer to in JAVA/C programming. Sketch encompasses the instructions which enables the Arduino to run.

SPI – SPI or serial peripheral interface which is used as communication protocol in short distance communication.

SRAM –   SRAM stands for Static Random Access Memory which implies that it does not have to be refreshed from time to time.

UART –   UART stands for Universal Asynchronous Receiver/Transmitter which is used to transform between serial and parallel data.

CHAPTER FOUR

ARDUINO IDE

Arduino IDE also known as Arduino Integrated Development Environment which is a text editor like program helps you to write Arduino code, verify the code and upload the same to Arduino board. When you save a file on Arduino, the file is called a sketch and a sketch is where you save all the computer code that you've written. Coding language that Arduino uses is very much like C ++ which is a common language in the world of computing. The code you learn to write for your Arduino will be very similar to code you write in any other computer language. All the basic concepts remain the same and it is really just a matter of learning a new dialect should you pursue other ventures and programming. The code that you will be writing is called human readable which implies that it will be organized for a human to follow. Part of the job is of the IDE is to take the human readable code that we write and then translate that into machine readable code that can be executed by the Arduino and that process is called compiling.

File extension for sketch is dot INO. It used to be dot PDE for older IDE version.

Arduino IDE can be divided into four parts:

1. File Menu – which can be seen in the top part of Arduino IDE

2. Toolbar

3. Code or Sketch window

4. Messages window

Buttons in toolbar offer easy access to functions which are commonly used. They are

1. Verify – This button verifies if the code is correct and error-free so that it could be uploaded to the Arduino board. Shortcut key for verify is Control R. It would show an error message at the bottom of the screen if there is any issue.

2. Stop – User may click on Stop button (to stop the operation of serial monitor) to receive snapshot of serial data to check it. Once stop button is clicked, other selected buttons will be un-highlighted .

3. New – User can click on New button to develop a blank sketch which helps you to key in the code. In order to create a blank sketch, Arduino needs name and location of the sketch. Giving default location

is considered as the best practice for giving location of the sketch.

4. Open – Open button offers a set of sketches which is saved in your sketchbook and also a set of sample sketches that you could use with multiple peripherals. For novice arduino users, this will definitely prove to be a valuable asset. User may edit the sample codes to suit their requirements.

5. Save – Save button helps to save the code that you have written inside the sketch window.

6. Upload – This button helps to upload the code that you have written inside the sketch window to Arduino. Short cut key for upload is control U. If you have got your Arduino hooked up to your computer with your USB cable, you'll see that 2 LED`s are going to blink really fast. The LEDs are T X and the R X  LED`s that stand for transmit and receive.

7. Monitor – It displays serial data which is sent from Arduino. It also helps to transmit serial data to Arduino. Short-cut key for serial data monitor is shift control N.

Baud rate is defined as the rate at which bits are send to or from the Arduino board. Default settings for baud rate is 9600 baud, user may change the baud settings if required. You may view error messages (if any) in red text at the bottom of IDE window. At the bottom left side, you may see a number which refers to the cursor location inside the code. If you navigate down the code lines, you could see that this number increases, this helps to pinpoint defects which are highlighted in the error message.

Best practices while dealing with Arduino IDE

1. Giving default location is considered as the best practice for giving location of the sketch

2. Make sure that you save the sketch before you upload it to Arduino board. This will help to retain code entered incase system hangs or IDE crashes.

3. Click on Verify/Complete button to confirm that your code is error-free.

IDE Menus

Now, let us learn about IDE Menus. The first menu is the Arduino menu. About Arduino displays the present version number and some additional details. Preference option enables user to change different IDE choices like sketchbook location. Quit Arduino helps to quit the IDE.

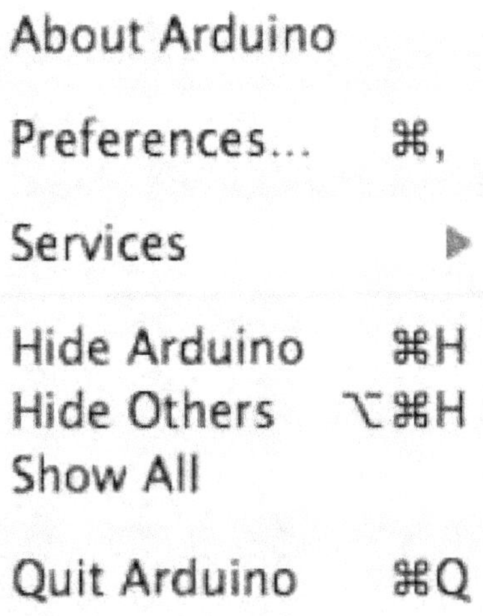

About Arduino

Preferences...   ⌘,

Services   ▶

Hide Arduino   ⌘H
Hide Others   ⌥⌘H
Show All

Quit Arduino   ⌘Q

Fig 4.1 Arduino Menu

File Menu

File Menu enables you to develop a new sketch , view sketches that you have saved in the sketchbook, save the sketch, upload the sketch and also to print out the code.

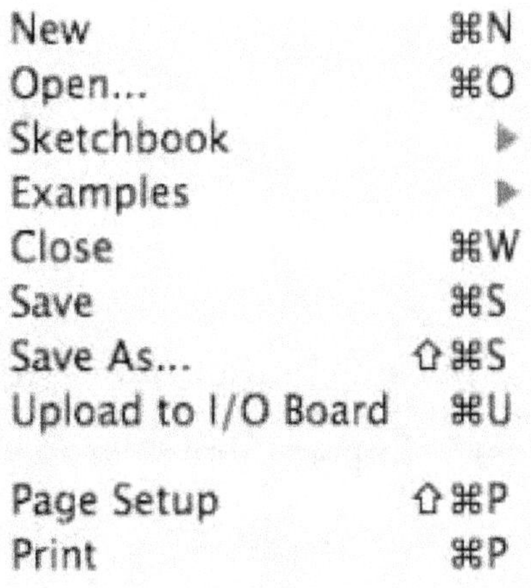

New   ⌘N
Open...   ⌘O
Sketchbook   ▶
Examples   ▶
Close   ⌘W
Save   ⌘S
Save As...   ⇧⌘S
Upload to I/O Board   ⌘U

Page Setup   ⇧⌘P
Print   ⌘P

Fig4.2 File Menu

Edit menu

Edit menu enables you to perform edit operations on the code – cut, copy and paste code sections. You can also increase or decrease the indents with the help of edit menu. This menu also helps to select the complete code or find specific codes or phrases inside the code.

Undo	⌘Z
Redo	⌘Y
Cut	⌘X
Copy	⌘C
Copy for Forum	⇧⌘C
Copy as HTML	⌥⌘C
Paste	⌘V
Select All	⌘A
Comment/Uncomment	⌘/
Increase Indent	⌘]
Decrease Indent	⌘[
Find...	⌘F
Find Next	⌘G

Fig 4.3 Edit Menu

Sketch menu

It encompasses Verify/Compile functions which helps to verify or compile the code. Import library feature displays list of available libraries saved inside libraries folder. Add File feature enables user to incorporate another source file to sketch which helps to separate bigger sketches into smaller units which can be then included in the main sketch.

```
Verify / Compile      ⌘R
Stop

Show Sketch Folder    ⌘K
Import Library...       ▶
Add File...
```

Fig 4.4 Sketch Menu

Tools menu

Tools Menu offers a galore of options. Auto format function renders your code crisp and nicer. Archive Sketch feature enables you to compress the sketch into a ZIP file format and asks you the location where you would like to store it. Burn Bootloader feature burns the Arduino Bootloader which is the piece of code on the chip which is conceived to make it compatible with IDE.

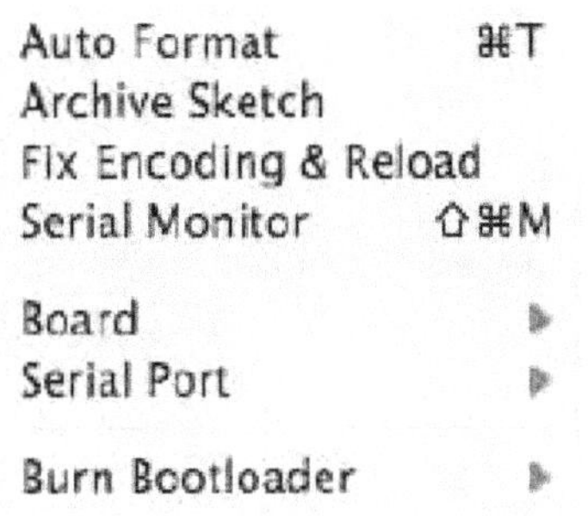

```
Auto Format           ⌘T
Archive Sketch
Fix Encoding & Reload
Serial Monitor        ⇧⌘M

Board                  ▶
Serial Port            ▶

Burn Bootloader        ▶
```

Fig 4.5 Tools Menu

Drop down menu

Drop down menu allows you to make new tabs, rename your tab, delete a tab and you can also navigate through the tabs down at the bottom.

Help menu

You can find valuable information pertaining to Arduino IDE and links to other useful pages.

CHAPTER FIVE

UNDERSTANDING ARDUINO SYNTAX

Syntax is to computer language as grammar is to written language. When we write a sentence, it might have a period, a semi colon or a dash and all those grammatical instruments to convey information to the reader. Computer programming languages also follow the same format and it is called as syntax.

Now, click on file →examples →basics and open up digital read serial. Let us learn step by step about the Arduino syntax.

```
/*
  DigitalReadSerial
  Reads a digital input on pin 2, prints the result to the seri

  This example code is in the public domain.
*/

// digital pin 2 has a pushbutton attached to it. Give it a na

int pushButton = 2;

// the setup routine runs once when you press reset:
```

Fig 5.1 Sample Code

In the above image, you could see two forward slashes. You could also see that all the words on the line following the forward slashes is grayed out. Comment tells the compiler to ignore that line of code. Comments allow you to explain to yourself or more importantly a future version of yourself what exactly you were doing when you wrote a piece of code.

Now, let us analyze int pushButton = 2;

Here, INT stands for integer. It has got a name push button and is assigned value as 2. If comment is not written, you would be wondering what the code means. Here, author has written explicitly that digital pin 2 has a push button attached to it give it a name.

Comments provide information to you when you forget what you wrote and it will be helpful to other users who view your code. Comments can be classified into two – single line comments and multi-line comments.

Semi colon is to grammar what a period is. The end of every statement of code need to have a semi colon that lets the compiler know that you are done with that statement of code. You will get an error if you forget a semi colon so every single line of code that you write other than function need to have a semi-colon .

There are two very common functions that you use in almost every Arduino program- they are called as set up and loop functions. Function basically encapsulates an extremely useful piece of code and then reduces it to a single keyword which allows you to rapidly implement that piece of code . So "set up" it is only one word but behind the scenes set up is a big function that does lots of stuff.

We put the opening and closing parentheses. For every opening parentheses, there should be closing parentheses. We put our opening curly bracket and now we can use all the power set up. Setup basically sets up the program. This is where you're going to initialize a pin as an input or output on your Arduino. You might set some baseline parameters for your program and start serial communication.

Now the loop is a function just like set up function. Loop is really the meat and potatoes of the function and the loop runs over and over again. Code gets executed from first line till it gets to the bottom and it gets to that curly closing bracket. This process repeats and that is why we call it a loop.

What void basically says is that this function does not return any information back to the program. So it's basically performing a function and that function in and of itself is all you need. You're not getting like a number back from set up.

CHAPTER SIX

PROGRAMMING EXPRESSIONS

Expressions are combinations of variables, constants, mathematical operators, logical operators etc.

Variables and data types

Variables are fundamental to all programming languages and the Arduino is no exception. Variables are programming tools that allow us to carry around information from one part of a program to another part of a program. Contents of the variable changes and variable stays as such. When you make a variable you need to specify what type of stuff that you're going to put inside the variable. In programming terminology, the stuff that you specify is called the data type. Some examples of data types are integers, characters and arrays. A variable will only hold what data type that you specify. The process of making a variable is called declaring or declaration. When you declare a variable there's two things that you need to tell the program. First one is data type and second is that you need to name the variable.

To declare a variable you need to write the type of contents it will hold (data type) followed by the name and there should be space in between.

For example, if you are going to write a program to blink an LED, we need to declare variable as

int led;

here int is data type, led is the variable and then it ends with a semi colon. When you declare variable in Arduino, you could notice is that the data type is a different color and that's because Arduino knows variable data types and gives them a special color to reduce confusion. When we specify data types what we're telling the compiler is to set aside much space for this information for this variable.

There are certain restrictions for variable naming. For example a variable name cannot have spaces or special characters in it. You may use upper and lowercase letters and numbers for naming the variable. However, variable name can't just be a number and your variable must start with a letter not a number. Variable names cannot be the name of Arduino keyword. It is good to give the variable name as descriptive of X function as possible.

You should always start your variables with lowercase letters and then if your variable name is two words connected together then you should capitalize the first letter of the second word and then if you have multiple words just keep capitalizing the first letter of the of the words as you go. For example, ledPin and holdPinValues.

Process of putting something in variable is initialization.

For example

Declaration is int led;

Initialization is led = 10;

We can couple together declaration and initialization as int led = 10;

There are several types of data types that you can declare. For example,

integer data type - an integer is a whole number and a whole number is just a number with no decimals (for example, 10 and 1500 are whole numbers but numbers like 1.5e is not a whole number). For Arduino an integer is a number from -32, 768 to + 32, 768.

boolean data type - it can take two values, either true or false

character data type - it can take values such as 'a' or 'X' etc.

long data type – this data type can be considered as an extension of integer data type and it could take values between -2157483648 to 2157483647.

Float data type – this data type can take values between -3.4028e+38 to 3.4028e+38.

Operators

Variables are manipulated by using operators. Operators can be broadly classified into four:

1. Assignment

2. Mathematical

3. Logical

4. Boolean

Assignment operator – it is also known as equal sign which computes the value in RHS and assigns the value with the variable in LHS.

Mathematical operator – they perform mathematical operations,

for example., + (addition), - (subtraction), * (multiplication), / (division)

Logical operator –

== equal to

! = not equal to

< less than

> greater than

< = less than or equal to

> = greater than or equal to

Boolean operator – It works on boolean data types

&& and

|| or

! not

Increment and decrement operators

Increment operator – Value of variable is incremented by 1

Decrement operator – Value of variable is decremented by 1

CHAPTER SEVEN

ARDUINO PROJECT CREATION PROCESS

Now, let us see what are the different steps involved in Arduino project creation.

Different steps involved in Arduino project creation are as follows:

1. Specify

2. Design

3. Prototype

4. Algorithm

5. Sketch

6. Compile and Upload

7. Test and Debug

Specify

Before you have to start with Arduino project creation, you need to be clear of the objective meaning, what is the nature and range of input given ? how the output would be and how it could be generated ? What process you need to go through to convert input into output ? Only when this phase is clear, you could proceed to next step

Design

Once you are clear with input type, output type and the transformation process, you may start working on the second phase which is design phase. You need to design circuit within the restrictions of the Arduino board to achieve this goal. You should be clear of the electronic components that you need to start working on the project. Electronic components may be resistors, capacitors, LEDs, sensors etc. You should also be aware of the specs of the components. You also need to have good understanding of the pins that are used as input, output or the pins which are not needed.

Prototype

Now, you have to create prototype of the circuit, this could be done either in breadboard or you can also make use of an online prototyping tool.

Algorithm

This part is often neglected by Arduino developers, but need to be given proper focus to ensure successful completion of the Arduino project.

Sketch

Sketch is analogous to code that we refer to in JAVA/C programming. Sketch encompasses the instructions which enables the Arduino to run. When you save a file on Arduino, the file is called a sketch and a sketch is where you save all the computer code that you've written.

Compile and upload

Compiler translates code that you have written into a form that Arduino microcontroller can perceive. Once the code is compiled it has to be uploaded.

Test and debug

This is one of the most time consuming parts of Arduino programming. If your code does not run, you need to debug the code, find the issue and fix it. You also need to check the circuit and ensure that it is proper.

CHAPTER EIGHT

DOWNLOAD AND INSTALL THE ARDUINO IDE

One of the absolute best things about the Arduino platform is easy it is to get started. The software that installs in your computer is completely free and its designs specifically for ease of use.

Go to your favorite browser and go to the Arduino homepage that's arduino.cc

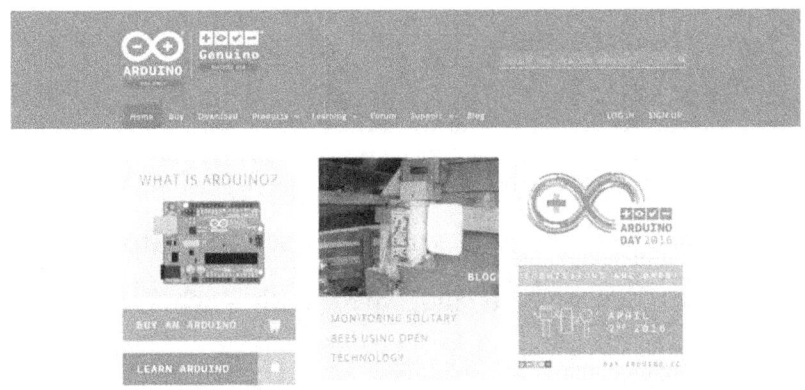

Fig 7.1 Arduino Homepage

Go to the download tab and find your operating system. We will install on the Mac for this tutorial.

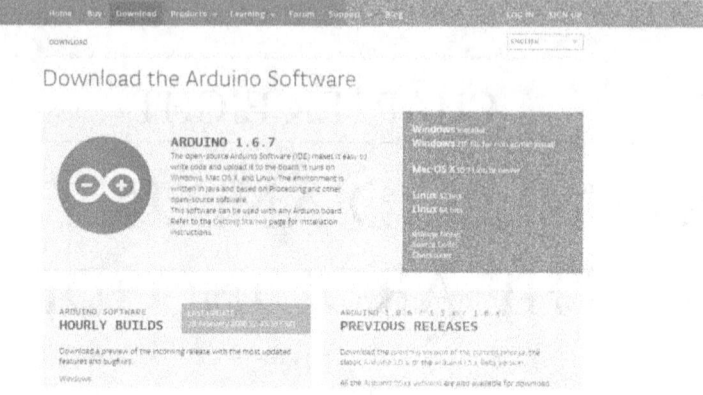

Fig 7.2 Arduino download

Now, you have to mention where you want to save it.

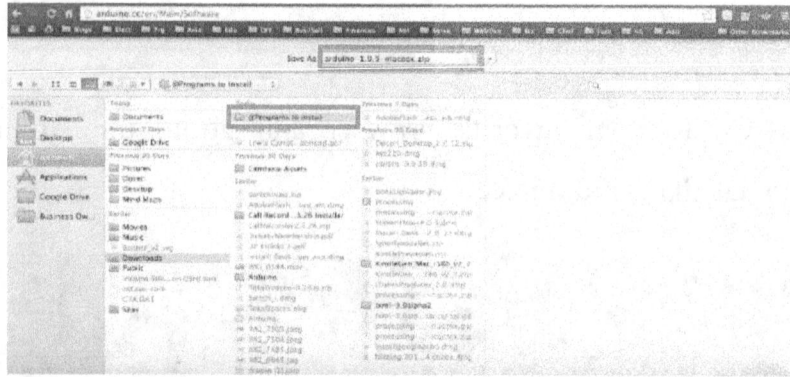

Fig 7.3 Arduino save

Go to My Finder and you could see that Arduino file is downloaded. Click on that and Open With Archive Utility and unzip the file.

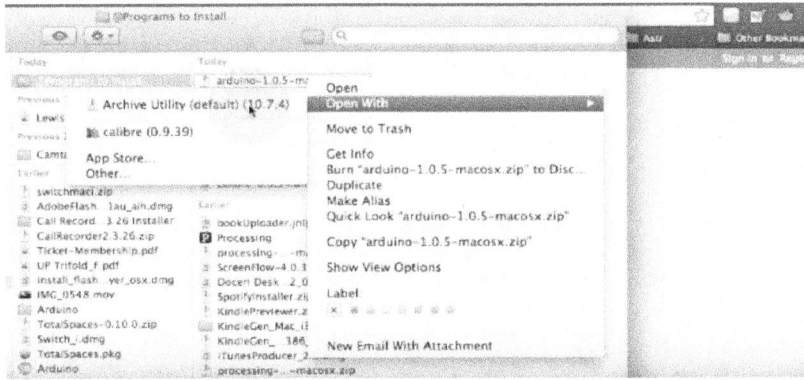

Fig 7.4 Arduino Unzipping

You can see Arduino now and drag it over the Applications and drop it there.

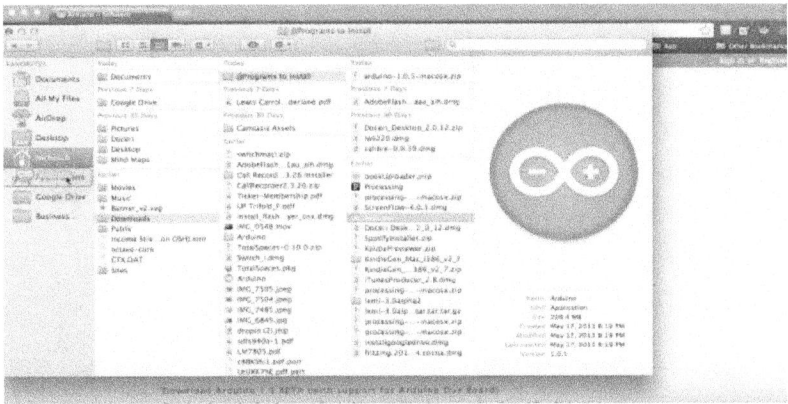

Fig 7.5 Arduino downloaded

Click on the applications and you could see the Arduino icon there.

Fig 7.6 Arduino icon

Launch Arduino IDE and select preferences

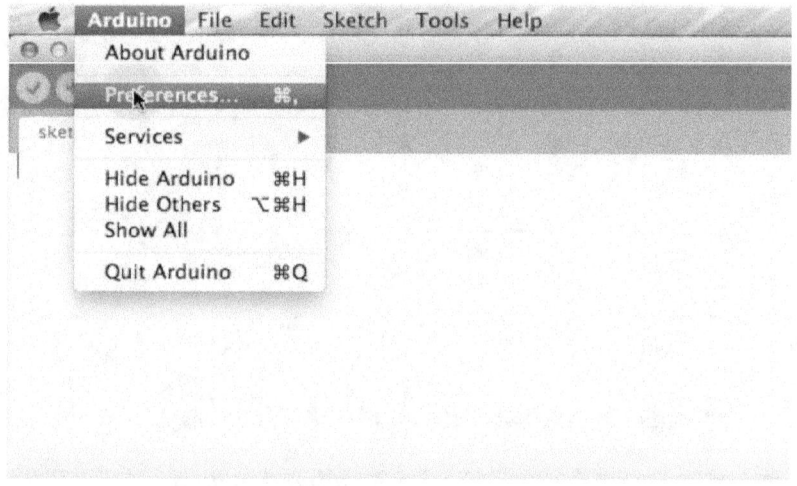

Fig 7.7 Arduino Preferences

Sketchbook location – This is where you want to store the sketches or the programs you will write for Arduino. You can change the sketchbook location by clicking on Browse button.

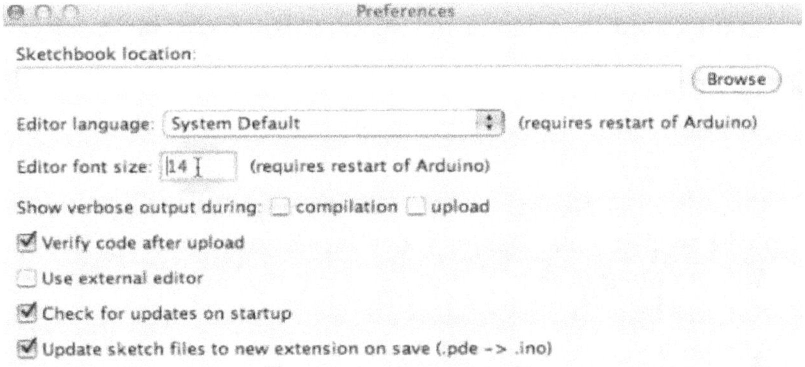

Fig 7.8 Arduino preferences window

Editor language – There is a galore of language options like Arabic, Danish, Dutch , French etc.

Editor font size – You can adjust font size. If you make font size too big, Sometimes your cursor would not match up with the lines.

Usually, verify code after upload option will not be checked. It is a good programming practice to click it.

Now go to menu and click tools and select Board.

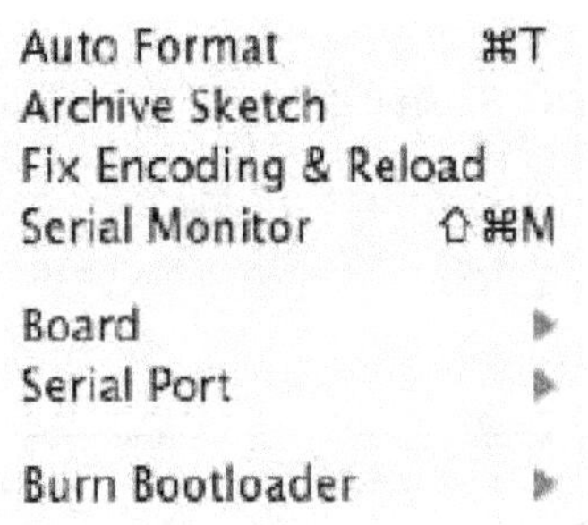

Fig 7.9 Board

You could now see a list of boards. If you have an UNO, select that option.

✓ Arduino Uno
Arduino Duemilanove or Nano w/ ATmega328
Arduino Diecimila, Duemilanove, or Nano w/ ATmega168
Arduino Mega 2560
Arduino Mega (ATmega1280)
Arduino Mini
Arduino Fio
Arduino BT w/ ATmega328
Arduino BT w/ ATmega168
LilyPad Arduino w/ ATmega328
LilyPad Arduino w/ ATmega168
Arduino Pro or Pro Mini (5V, 16 MHz) w/ ATmega328
Arduino Pro or Pro Mini (5V, 16 MHz) w/ ATmega168
Arduino Pro or Pro Mini (3.3V, 8 MHz) w/ ATmega328
Arduino Pro or Pro Mini (3.3V, 8 MHz) w/ ATmega168
Arduino NG or older w/ ATmega168
Arduino NG or older w/ ATmega8

Fig 7.10 List of Arduino boards

Now, select the Tools menu and click serial port and select the suitable port from the Arduino list. Now, we may upload a sketch for further study.

Uploading your first sketch

Select the File Menu and click Examples.

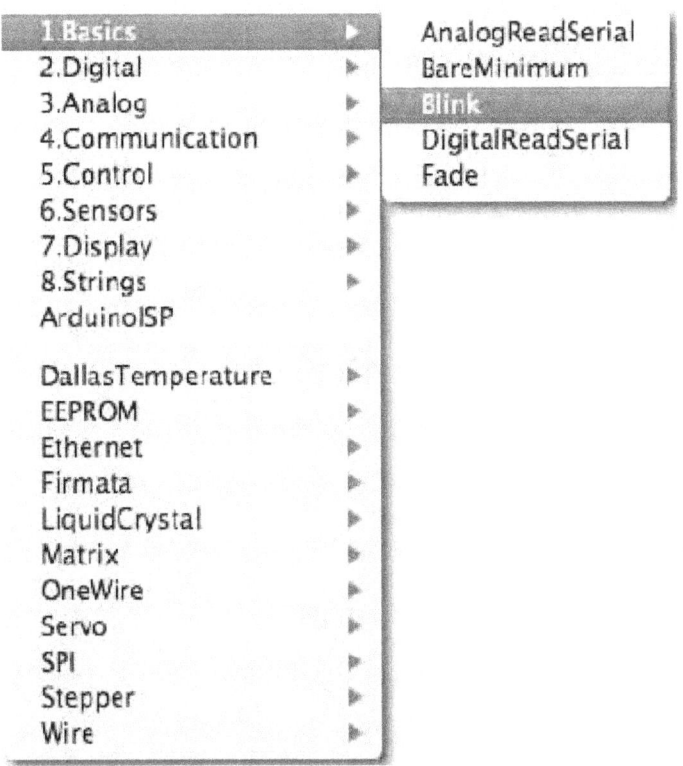

Fig 7.11 Uploading first sketch

You may see a list of examples here. Let us start with a simple program. Select Basics and click on blink option. Blink Sketch will be now loaded into Arduino IDE. Now, click on upload option and have a glance at the Arduino. You could see that RX and lights have started to blink which implies that data is send from your system to Arduino board. When you have successfully uploaded the sketch, you will get a message "Done uploading" in the IDE status bar. Now, RX and TX LEDs will stop flashing.

In a short while, you may see that Pin 13 starts to flash on and off at one second intervals. If this happens, you can be sure that you have successfully uploaded the sketch.

Code required to make an LED blink

```
void setup() {

        pinMode(13,OUTPUT);

    }

    void loop() {

        digitalWrite(13, HIGH);

        delay(1000);

        digitalWrite(13, LOW);

        delay(1000);

    }
```

Code explanation

Here, you are assigning digital pin 13 as output pin. Now, we give a loop which switches pin 13 onto HIGH or LOW for 1 second. Delay time we have given is in milliseconds, which equals to 1000 milliseconds or 1 second.

Fig 7.12 LED blinking

CHAPTER NINE

HARDWARE OVERVIEW

Arduino is an open source hardware which means that all the design files for this board have been made public. So anybody can go and make clones of their Arduino. Now, let us learn in depth about Arduino Hardware. Pins in Arduino UNO can be classified into 3 categories – Power, analog in and digital.

Digital pin header is a plastic line that has got a bunch of holes in it and has some numbers next to it. They are numbered from 0 up to 13 from right to left, which means that there is actually 14 little hole that are numbered which give you access to the pins on the chip right there . These pins can either act as inputs so they can read levels of voltage from devices or they can act as output so they could apply voltage like 5 volts.

Digital pins 3,5,6,9,10 and 11 can use pulse with modulation which implies that they can adjust the amount of voltage they apply between 0 and 5 volts. Pins 0 and 1 have a TX and an RX so that stands for transmit and receive . Now if you are transmitting data with your computer Pins 0 and 1 will be utilized. If your project involves a lot of communications, for example serial communications with your Arduino, it is better not to use these transmit, receive pins too much as it might affect how your program operates. Pins 2 and 3 are used as external interrupts which helps to trigger an action if the system perceives external events. Pins 10,11,12,13 are used as SPI pins.

Transmit and receive LED's are also embedded on the board. These LEDs will blink anytime you are sending or receiving data. When you load a sketch or load a program onto Arduino board those LEDs will blink. These LEDs are really good for troubleshooting. If you think you are sending data to your arduino board but the received LED isn't blinking, then it means that you are probably not sending anything.

Analog pin headers are marked 0 to 5 from left to right which means that there is 6 analog pins. Analog pins give you access to analog to digital converter. Analog to digital converter (ADC) enables you to take analog signals and convert them into digital signals. Analog to digital converter takes the infinite amount of variation in an analog signal and it digitizes into small steps. This is really beneficial as most of the sensors that you use will provide you analog information. Analog pins could be used like the digital pin headers.

3.3 V stands for 3.3 volts and 5V stands for 5 volts. Now if you have your Arduino connected to either a power supply like battery power supply or you have hooked up to your computer with a USB port, you can get 5 volts or 3.3 volts respectively from those pins. If you connect a wire into that 5 volts hole, and measure the end of that wire using multimeter it will be 5 volt. Similarly, you may measure for 3.3 volts also. You will be using 5 volt pin frequently compared to 3.3 V for all the Arduino sketches.

When you click on the reset button, Arduino will start over at the beginning of the program. However, reset will not erase anything off the board, but it performs reboot in a quick manner. There is also a reset pin adjacent to the 3.3 and 5 old pin headers. You can reset the board by applying zero volts to the reset pin. There are 3 ground pins on the Arduino board, 2 ground pins are located on the bottom next to the 3.3 and 5 volt pin headers and there is one more ground pin on the top, next to digital pin 13. Ground pin gives you access to the lowest voltage on the Arduino board.

There`s also a power on LED. When you have the power applied to Arduino board either from your USB to your computer or with an external battery power connect, this light will be turned on. Finally if you do have an external battery connected you`ll see it is this port here where you would connect that battery power.

Arduino toolbox

All you need to get started with Arduino are:

- Arduino

- Mini breadboard and jumpers (Breadboards are extremely useful for prototyping)

- LEDs and resistors

- Battery and battery connector

- LDR or light-dependent resistor

- DC motor

- Piezo buzzer and potentiometer

Common components

Let us discuss about some of the common electronic components

Resistors

Resistor is a passive electronic component which helps to limit electric flow in the circuit. Electrical properties of a resistor is measured by means of its resistance. Resistance of a resistor can be measured with the help of color-coded bands provided on the resistor. Resistance is measured in ohms. Potentiometers, adjustable resistors and resistance decade boxes are some of the examples of variable resistors.

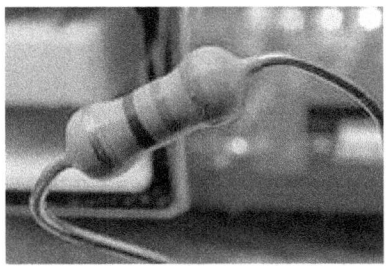

Fig 8.1 Resistors

Capacitors

They are also passive electronic component like resistor and helps to store electrical energy when current flows through it. Capacitors are associated with a property known as capacitance which measures the energy storage capabilities. Capacitance is measured in farads. Capacitor encompasses two parallel plates made of conductive material and includes a dielectric material between them. Capacitors find their relevance in a multitude of applications like energy storage, filters and noise suppression. They are also used in motor starters, tuning circuits, oscillators etc.

Fig 8.2 Capacitors

Diode

It is an electronic component with two terminals and allows flow of current in only one direction. Most of the diodes are made of silicon. Diodes are of different types like PN junction diode, tunnel diode, Varactor diode, GUNN diodes, Zener diodes, photodiodes etc. Diodes are used in different applications like radio demodulation, power conversion, clipping circuits, clamping circuits, over voltage protection etc.

Fig 8.3 Diode

LED

LED stands for light emitting diode which is a p-n junction diode and emits light when it is triggered. Working principle of LED is known as electroluminescence – when LED is triggered by proper voltage between its leads, electrons recombine with holes

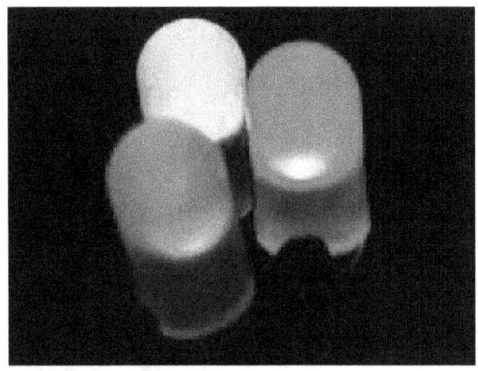

to release energy.

Fig 8.4 LED

Transistor

Transistor stands for transfer of resistance and they are used to amplify or switch electronic signals. Most common type of transistor is Bipolar junction transistor (BJT transistor) which is classified into two types pnp transistor and npn transistor. Transistors find use in electronic circuits as a switch and an amplifier. Transistor has three pins – emitter, base and collector.

Fig 8.5 Transistor

Relay

It is an electrically operated switch and is available in different forms like latching relay, reed relay, solid-state relay, vacuum relays etc. They are used to regulate either a high power or high voltage circuit with that of a low power circuit.

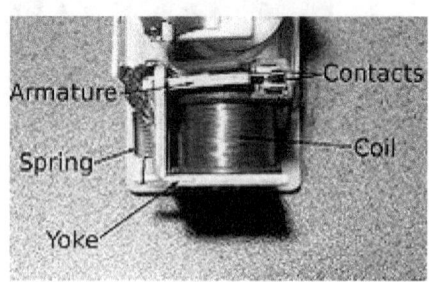

Fig 8.6 Relays

Shields

Shield is a PCB(printed circuit board) which encompasses multiple built-in components which helps to accomplish a specific task. Shields can be mounted into Arduino board directly without the need of wiring or soldering. Shield has to be aligned with Arduino board and apply pressure to mount shield into Arduino Board. Shields are of different types like Ethernet and Wifi networking, Bluetooth, GSM, GPS, sensors etc. Shields are also available for prototyping which renders it easy to retain the best circuit designs. For novice Arduino users, Shields are a great blessing since they don't need any tools to incorporate components to Arduino Board.

Fig 8.7 Shields

Breakouts

They are normally small circuit boards designed around an IC to offer particular functionality. It encompasses auxiliary circuitry like LEDs, resistors and capacitors for attaching the breakout unit to different modules or to Arduino board. You may find the same functionality in breakouts as shields. However, breakout is smaller compared to shields and go well with other boards as well. Breakouts are attached to board with the help of jumper wires and breadboard. At times, you may also need soldering to connect to the board or modules.

Fig 8.8 Breakouts

CHAPTER TEN

ARDUINO COMMANDS

A set of commands are used to interact with the Arduino Board.

pinMode(*pin,mode*) – Here, pin need to be an integer value. Modes can take any of the following values – INPUT, OUTPUT and INPUT_PULLUP

digitalWrite(*pin,value*) – Here, pin need to be an integer value. Value can be either high or low.

digitalWrite(*pin*) – Here, pin need to be an integer value. Value returned can be either high or low.

digitalRead : Syntax int i = digitalRead(pin);

Here, variable i is set to either of two values – high or low.

analogRead(*pin*) – This command returns analog voltage reading at the pin.

analogWrite(*pin, duty cycle)* – This command is used to write PWM value to pin. Duty cycle can take value between 0 and 255, where 0 refers to off state and 255 refers to on state. Pins can be either of 3,5,6,9,10 and 11.

analogReference(type) – Here, user may select from any of the following options

1. DEFAULT voltage level is 5 volts and 3.3 volts respectively for 5V and 3.3 V arduino board respectively.

2. INTERNAL is an in-built reference which differs with processor type.

3. INTERNAL1V1 is an in-built 1.1 V reference only in Mega Arduino board.

4. INTERNAL2V56 is an in-built 2.56 V reference only in Mega Arduino board.

5. EXTERNAL – this implies that you could make use of voltage applied to AREF pin as reference voltage.

pulseIn : Syntax i = pulseIn(pin, value) – This command retrieves the period in microseconds of the next high or low pulse in the specified pin number. Value can be either high or low.

tone : Syntax tone(pin, frequency, duration) – This command makes the pin oscillate at the mentioned frequency for specified duration. Example – tone(8,500,1500).

noTone – This command cuts short the playing of tone which was in progress.

millis – This command returns the number of milliseconds after the last reset.

micros – This command returns the number of microseconds after the last reset.

delay(millisecond) – This command returns the delays in milliseconds.

delay(microsecond) – This command returns the delays in microseconds.

Random functions

We have seen above timing functions like millis(), micros(),delay(millisecond) and delay(microsecond). These functions work in a rather straightforward manner. For example, it explicitly specifies the time duration for there is delay, may be 1000 millisecond or 5000 microsecond. Random functions work in a random manner, where time delay could vary between the minimum and maximum time limits that you specify. Random() takes the below mentioned syntax

random(min,max)

Time delay can range anywhere between the min and max parameters mentioned in the random function. Here, min is the minimum value of time delay whereas max is the maximum value of time delay. If no value is mentioned in the random function it could take any value which lies between - 2,147,483,648 to 2,147,483,647.

CHAPTER ELEVEN

ADVANCED PROGRAMMING CONCEPTS

Control Statements

Code written inside the loop is executed from the top to bottom and once commands are executed it will return to top of the loop and starts execution again. Program flow in Arduino programming is based on two types of statements – conditional statements and iterative statements.

If statement

It is the simplest of the control structures and is used widely in the decision making process. Syntax for if statement is as below:

If(condition) {

Statements

}

If the condition given is matched, then control is driven to the loop. Else, loop will not be executed.

If statement can be extended as if-else statement as below:

If (expression)

{

Statements

}

Else

{

Statements

}

For statement

It is an iterative statement which helps to execute the code provided till it violates the loop condition.

Syntax for the statement is as below

For(declaration; condition; increment) {

Statements

}

For example for (i=0; i <10, i++){

Statements

}

Here, initially variable is declared as 0 and then the variable is checked against the condition. Value of variable is incremented by 1. Control goes into the loop and statements are executed. This process repeats until the condition is violated.

While

While statement executes the loop until the condition remains true.

Syntax

While (condition)

{

Statements

}

Do while

This loop is analogous to While loop, however the testing condition is provided at the end. This ensures that the loop is executed atleast once.

Syntax

Do {

Statements

} while (condition)

Switch statement

Switch statement is similar to if else statement and is well structured. Switch statement can be used only if you would like to test one variable. If none of the conditions are valid, default statement will be executed.

Syntax for switch statement is like below:

switch(expression)

    {

        case constant-expression1:   statements1;

        [case constant-expression2:   statements2;]

        [case constant-expression3:   statements3;]

        [default : statements4;]

    }

Break

This keyword helps to exit from a loop or a switch and control is transferred to the first statement which comes after the loop/switch.

Interrupts

Interrupts enable microcontroller to respond to events without checking continuously to verify if anything has changed. You may also make use of timer-generated interrupts.

Hardware interrupts

Let us see a sample code below:

```
void loop

{

if ( digitalRead(input pin) == LOW)

{

// do something

}

}
```

In the above mentioned code, we check continuously to check input pin and the instance it turns low, processor need to perform the events that is mentioned in the loop. Hardware interrupts help to receive inputs (even if they are shorter pulses for minute duration). In Arduino UNO, there are only two pins which can be used as hardware interrupts.

Resistor retains the interrupt pin (D2) HIGH until the button on the switch is pushed. Once D2 is grounded it will turn LOW.

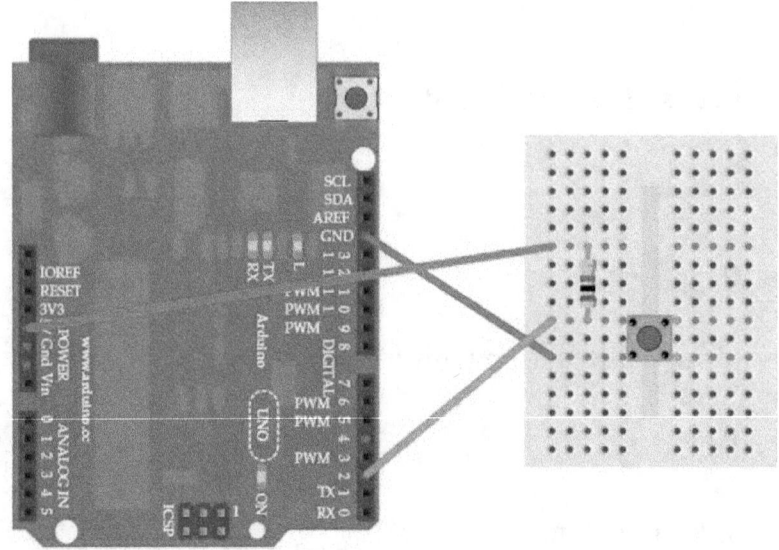

Fig 10.1

Sketch

```
int ledPin  = 13;

void setup()

{

pinMode (ledPin, OUTPUT);

attachInterrupt ( 0, stuffHappened, FALLING);

}

void loop ()
```

```
{

}

void stuffHappened ()

{

digitalWrite(ledPin, HIGH);

}
```

Code explanation:

pinMode (ledPin, OUTPUT); - sets the LED pin to be an output

attachInterrupt (0, stuffHappened, FALLING); - here the argument 0 refers to interrupt number (In Arduino UNO, there are only 2 interrupt pins, interrupt 0 is pin D2 and interrupt 1 is pin D3. Argument stuffHappened is the function name that has to be used when interrupt happens. This function is known as Interrupt Service Routine or ISR which is analogous to void function in that it does not take parameters nor do they return anything. The last parameter that we have given is a constant, FALLING. This implies that interrupt would result in interrupt service routine only when D2 falls from HIGH to LOW. (Resistor retains the interrupt pin (D2) HIGH until the button on the switch is pushed. Once D2 is grounded it will turn LOW. We need to call the interrupt only when it is grounded). In the sample code, we have left the loop function as blank. If we need to perform any action in the case of an interrupt, we need to mention that in the loop. Usually, we would write code to light up LED in case of interrupt.

Arduino Library

Arduino library is a set of functions which helps to accomplish particular tasks which renders life effortless for you. We have seen functions like digitalWrite() and delay() which is included in Arduino library. Every library needed to be incorporated in the sketch when you would like to include a particular function. An example for arduino library is LiquidCrystal library which enables you to communicate with LCD displays if it is incorporated in the sketch. Some of the Arduino libraries work as a stand-alone library, whereas other Arduino libraries has to be utilized with extra electronic components which is also known as shield.

Arduino library can be broadly classified into three:

1. Core

2. Standard

3. Contributed

Using libraries

Arduino programmer need to explicitly mention the complier about Arduino library before they are added. # include pre-processor directive has to be added to mention file name of Arduino library. Pre-processor directive includes instructions which enables the compiler to add relevant function in the Arduino sketch.

Syntax is as follows:

include < LibraryName.h>

You need to replace LibraryName with the file name of Arduino library. For example, if you would like to communicate with LCD displays, you need to write as

include < LiquidCrystal.h>

You may also add library by navigating through Sketch menu and then click on Import library. You need to select library name from the menu. Once you have added the library, you may need to create either a library instance or perform library initialization. For example, library instance can be created as below:

LiquidCrystal lcd(5,6,7,8,9,10);

Here, the name of library instance is lcd and we have assigned pin numbers 5,6,7,8,9 and 10. If you would like to add multiple library instances, you may do so by creating library instances as lcd1, lcd2 etc.

Library Initialization

Now, let us see how we could initialize Arduino library. We need to define size of LCD that we are going to implement. This can be done using begin(cols, rows)

lcd.begin(16,2);

Here, we have created an instance known as lcd and Arduino function begin. Two parameters are passed in this arduino function which describes that LCD display is 16 characters wide and 2 characters tall or a 16*2 display.

print()

This function is used to transmit information to liquid crystal display unit. Syntax for print() can be either of the following:

print(data)

print(data, BASE)

Data could be anything that you would like to see in the LCD screen as output. It has to be provided in double quotes. An example would be like

lcd.print(" Success");

clear()

This function helps to clear display content and place cursor in top left side of the LCD display. clear() is analogous to void() and does not need any argument. This function helps to write new line of code without the need of overwriting data.

setCursor()

This function helps to change the cursor position to a new position without clearing the display content. Syntax is

setCursor(col,row);

write()

This function helps to transmit character to LCD display. Syntax is

write(data)

createChar()

This function helps to create a symbol for example smiley face. To accomplish this, one need to describe a bitmap array which matches up with every dot for the character. createChar() is then used to transmit the bitmap array to the relevant display memory address.

Syntax for createChar() is createChar(number,data);

scrollDisplayLeft()

This function enables the display to be shifted one character to the left.

scrollDisplayRight()

This function enables the display to be shifted one character to the right.

Servo library

This library can regulate upto a maximum of twelve servos in the Arduino UNO board. An instance of servo library can be created as below

Servo servo1; (Servo library name has to be provided followed by the instance of Arduino sketch here it is servo1).

attach()

This function has to be used to use servos, syntax for attach() can be either of the following

name.attach(pin) or

name.attach(pin,min,max)

In some cases, arduino will have minimum and maximum rotation angles which has to be provided. (Note: Rotation angle is mentioned in microseconds. Default minimum is 544 for 0 degree whereas maximum is 2400 for 180 degree).

write()

Write() helps the Servo to move and the syntax for write() is

name.write(angle)

Stepper Library

Stepper motor offers positional accuracy and could rotate for the complete 360 degree. As the name implies, stepper motor rotates in a specific number of degrees or steps. Rotation angle per step can vary from one stepper motor to another. An instance of stepper library can be created as follows:

Stepper instancename(steps, pin1, pin2)

setSpeed()

This function helps to set motor rotational speed. Syntax for setSpeed() is as below

name.setSpeed(rpm)

Here, we provides motor speed in terms of rotation per minute. (Note: To ensure that stepper motor runs properly, user need to select suitable speed for the motor. If the speed is too high, it can result in missed steps or sporadic operation of the motor.

step()

Motor can be moved by using step(). Syntax for step() is as follows

name.step(steps)

Parameter that we pass to step() is the count of steps through which the stepper motor need to move

SD library

SD stands for Secure digital flash memory which helps to write to an SD memory card.

File

File instance is created with the help of file and takes the below syntax

File Name

SD.begin()

SD library is initialized with the help of SD.begin().It will default to CS or chip select pin and takes the below syntax

SD.begin(csPin)

SD.open()

This function helps to write data to Arduino file and it takes the below syntax

SD.open(filename,mode)

Here, filename is the name of the file that you would like to access. Make sure to give short file names as this function works in a 8.3 format, which means file name could be character string with maximum of 8 characters and file extension is 3character extension like .doc. Second parameter, viz. mode can take two values either FILE_READ or FILE_WRITE. Default mode of SD.open() is FILE_READ which helps to open the file quoted in filename. FILE_WRITE helps to write data into the filename mentioned.

close()

This function helps to close the current file and takes the below syntax

name.close();

write()

This function helps to write into the SD card and takes the below syntax

name.write (data)

Name is the instance of file class and data is anything that you would like to input into the file. It could assume different data types like char, string etc.

available()

This function helps to know the number of bytes present in a specific file and takes the below mentioned syntax

name.available()

read()

This function is used to extract information from a specific file and takes the below mentioned syntax

name.read()

Arrays

Arrays are list of variables and includes multiple elements with same data type. Elements in an array can be accessed with the help of index number.

Array declaration

Array declaration is similar to variable declaration. For example,

int sampleArray[3];

In the above array declaration, three distinct variables are referred in the Arduino sketch as sampleArray[0], sampleArray[1] and sampleArray[2].

When we declare array, Arduino compiler will reserve a block of memory for our use. For array declaration we need three things –

array data type – It is int in the example

array name- It is sampleArray in the example

index – [3] in the example . Index refers to the count of variables in the array.

Assigning values

First element in the array starts at position 0 and the last element is at an index which is one less than array size.

int ledPins[3];

ledPin[0] = 8;

ledPin[1] = 9;

ledPin[2] = 10;

We can assign values like below also.

int ledPins[] = { 8,9,10};

Character arrays

In the above discussion, we dealt with integer arrays. Now, let us learn about character arrays.

Char Sample[] = " Good morning";

Multidimensional arrays

Usually, we work with one-dimensional arrays. However, we may need to work with multidimensional arrays at times. For example

int SampleArray[2][3];

The above mentioned array will contain total of 2*3 = 6 elements. Elements are as below:

SampleArray[0][0];

SampleArray[0][1];

SampleArray[0][2];

SampleArray[1][0];

SampleArray[1][1];

SampleArray[1][2];

Strings

Strings are not commonly used in Arduino environment compared to other programming languages like C, C++ etc. This is because, there is no need to represent String functions in Arduino, if at all used it is in the form of Serial.println command which is used to debug a buggy sketch.

Imagine, you are going to define a string constant as below

Char message [] = " Good day" ;

Here, you are defining a char array that is 9 characters long (not 8 characters long as it also includes a terminating character of 0 to denote the end of the string. Every character letter, number or symbol is assigned a code known as ASCII value.

String constant can also be defined as below

Char *message = " Good day" ;

In the above mentioned syntax, declares message as a pointer to a character.

Finding string length

String length can be found out using strlen function. It counts the number of characters in the array excluding the null that denotes the end of the string.

For example, strlen("hello") returns the number 5.

Some of the string functions

[]

char message = String("hello")[0] – this functions returns value h

Trim

String s = "   hello   ";

s.trim();

This function helps to remove space characters on either side of hello and returns hello (with no space).

toInt

This function converts string function into an int or long

StringReplace function

This function enables user to replace all instances of a given character with another character.

StringRemove function

This function enables users to remove specific part of a string. This function can take either one argument or two arguments. If one argument only is passed, string from that part of the index to end is removed. If two arguments are passed, string from the index of first argument to index of second argument is removed.

startsWith and endsWith

This function helps to check if a string starts or ends with a particular substring.

String comparison operators

There are multiple string comparison operators like ==, !=, >,<,>+=,<= etc. which helps to make alphabetic comparisons between strings. These functions are really helpful to perform sorting and alphabetizing.

String appending operators

These operators help to append things to strings. You may use StringAdditionOperator, += operator and concat() to append things to strings.

String case change functions

As the name implies, these functions help to change the case of the string either from upper to lower or lower to upper.

CHAPTER TWELVE

SAMPLE CODES FOR STUDY

Line tracking:

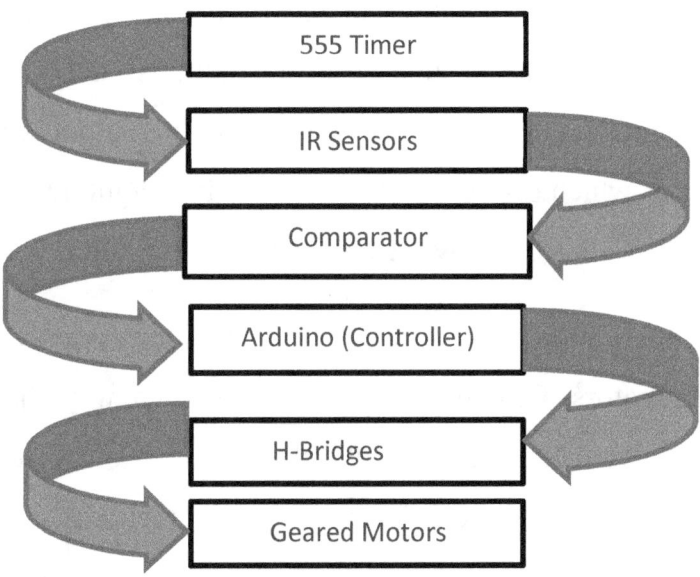

Fig 11.1 Block Diagram of Line Tracking

In line tracking robot two motors are used which are controlling two wheels. It has four infra-red sensors placed at the bottom for the detection of black lines. When two middle sensors detect the black color the output from the receiver is fed into comparator

(LM339). The comparator compares the output of the sensor to the reference voltage which is set as 5V to give its output.

The output of the LM339 is high when it receives input from the IR sensor. A very easy and simple logic is applied in line tracking. A black color absorbs radiations where as white or bright color reflects radiations. Four pairs of transmitter (Tx) and receiver (Rx) are used.

For detection of line, sensors are arranged in such a way that sensors face the ground. Sensors output is analog and its value depends on the amount of reflected light. Output is fed into comparator which gives logic 0's or 1's at its output. Direction of the wheels is controlled by the Arduino.

The rotation of the wheels depends on the output values of the comparators. Circuit is designed in such a way that when sensors are on black line it reads '0' and when on white line it reads '1'. There are three different cases of direction:

1. Straight

2. Right curve

3. Left curve

Straight Direction

When robot is moving in a straight direction, two middle sensors are working and give low response and remaining two sensors give high response. Arrangement is made in such a way that middle sensors are always online. As the color of the line is black, it will not emit radiations back and response will be always low. Remaining two sensors response will be high as they will be on white surface.

Right Curve

When robot is at the right curve on the line, response changes i.e. the response of the two sensors on the right side becomes low as sensor faces black line and remaining sensors response becomes high. Due to this change, right wheel is held and left moves freely until middle sensors come back to normal position.

Left Curve

When robot is at the left curve on a line, response changes i.e. the response of the two sensors on the left side becomes low as sensor faces black line and remaining sensors response becomes high. Due to this change left wheel is held and right moves freely until middle sensors come back to normal position.

Schematic of H-Bridges for controlling Motors:

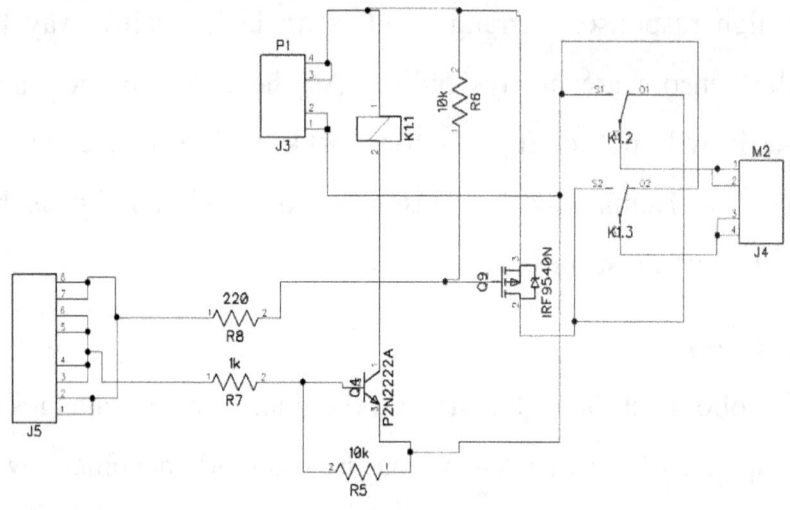

Fig 11.2 Schematic of H-Bridges for controlling Motors

Same Schematics will be used for another motor.

Code for Line Tracking:

void setup()

      {

      // put your setup code here, to run once:

      pinMode(10,OUTPUT);//PWM FOR MOTOR 1

      pinMode(7,OUTPUT);//RELAY   FOR   MOTOR   1
(Direction)

      pinMode(8,OUTPUT);//PWM FOR MOTOR 2

```
pinMode(11,OUTPUT);//RELAY FOR MOTOR 2
```
(Direction)

```
pinMode(A0,INPUT);//Sensor most left    XOOO

pinMode(A1,INPUT);//Sensor  left         OXOO

pinMode(A2,INPUT);//Sensor right         OOXO

pinMode(A3,INPUT);//Sensor most right   OOOX

}
```

//A0,A1,A2,A3 are analog pins of Arduino. The value received by sensor becomes the input of Arduino. Arduino pin 10 is connected to enable pin of motor1 whose output will activate and deactivate motor1. Arduino pin 8 is connected to enable pin of motor2 whose output will activate and deactivate motor2.Ardunio pin 7 and pin 11 are connected to relay which are further connected to motor1and motor2 respectively via relays. Its output will move motor in clockwise and counter clockwise direction. These pins can be changed to any digital pin of Arduino but the respective pin should be mentioned in code.

```
void loop() {

int s1,s2,s3,s4;   //Sensors==4 Sensors are used

s1=analogRead(A0);//Reading  from  analog  port  A0
```
//Actually Sensor1 Value

s2=analogRead(A1); //Reading from analog port A1 //Actually Sensor2 Value

s3=analogRead(A2); //Reading from analog port A2 //Actually Sensor3 Value

s4=analogRead(A3); //Reading from analog port A3 //Actually Sensor4 Value

//Sensors output is feed into Arduino which becomes its input. Setting threshold value by checking value of sensors. Each IR sensor has different output values when light is reflected .for example your sensor output is 2volts.Arduino ADC is 10 bit and reference voltage is 5v. (1024/5)*2v=approx. 400.This converted value is used in code.

if(s1>400 & s2<400 & s3<400 &s4>400) // White,Black,Black,White WBBW

//In this if condition sensor s1 and s4 are on white line and sensor s2 and s3 are on black line. As our robot will track black line, that's why motor is not changing direction. Relay is active logic low and thus motor moving in a straight line. In all the if statements given below depending upon the value of sensors it will either turn right, left or move in a straight line.

{

digitalWrite(7,LOW);//Relay of motor1

```
digitalWrite(10,HIGH);//PWM1

digitalWrite(8,LOW);// Relay of motor2

digitalWrite(11,HIGH);//PWM2
}
```

if(s1<400 & s2<400 & s3>400 &s4>400)// Black,Black,White,Black BBWW

// In this if condition sensor s3 and s4 are on white line and sensor s1 and s2 are on black line. As our robot will track black line, that's why robot will change direction. One Relay is active logic low and other will go active logic high thus robot will turn right.

```
{

do

{

digitalWrite(7,LOW);// Relay of motor1

digitalWrite(10,HIGH);//PWM1

digitalWrite(8,HIGH);// Relay of motor2

digitalWrite(11,LOW);//PWM2

}while(s1>400 & s2<400 & s3<400 &s4>400);
```

}

if(s1<400 & s2>400 & s3>400 &s4>400)   //
BlackWhiteWhiteWhite

///// In this if condition sensor s2,s3,s4 are on white line and sensor s1 are on black line. As our robot will track black line, that's why robot will change direction. One Relay is active logic low and other will go active logic high thus robot will turn right.

```
{

do

{

digitalWrite(7,LOW);// Relay of motor1

digitalWrite(10,HIGH);//PWM1

digitalWrite(8,HIGH);// Relay of motor2

digitalWrite(11,LOW);//PWM2

}while(s1>400 & s2<400 & s3<400 &s4>400);

}
```

if(s1>400 & s2>400 & s3<400 &s4<400)   //
White,White,Black,Black WWBB

//// In this if condition sensor s1 and s2 are on white line and sensor s3 and s4 are on black line. As our robot will track black line, that's why robot will change direction. One Relay is active logic low and other will go active logic high thus robot will turn left.

```
{
  do
  {
    digitalWrite(7,HIGH);// Relay of motor1
    digitalWrite(10,LOW);//PWM1
    digitalWrite(8,LOW);// Relay of motor2
    digitalWrite(11,HIGH);//PWM2
  }
  while(s1>400 & s2<400 & s3<400 &s4>400);
}
if(s1>400 & s2>400 & s3>400 & s4<400) //
White,White,White,Black WWWB
```

//// In this if condition sensor s1, s2 and s3 are on white line and sensor s4 are on black line. As our robot will track black line, that's why robot will change direction. One Relay is active logic low and other will go active logic high thus robot will turn left.

```
{

   do

   {

      digitalWrite(7,HIGH);// Relay of motor1

      digitalWrite(10,LOW);//E1

      digitalWrite(8,LOW);// Relay of motor2

      digitalWrite(11,HIGH);//E2

   }

   while(s1>400 & s2<400 & s3<400 &s4>400);

}

   if(s1>400 & s2>400 & s3>400 & s4>400) //
```

White, White, White, White WWWW

///// In this if condition sensors1, s2, s3 and s4 are on white line. As our robot will track black line, that's why robot will not change direction. It will stop moving.

```
{

   do
```

```
{

digitalWrite(7,LOW);// Relay of motor1

digitalWrite(10,LOW);//PWM1

digitalWrite(8,LOW);// Relay of motor2

digitalWrite(11,LOW);//PWM2

}

while(s1>400 & s2<400 & s3<400 &s4>400);

}

if(s1<400 & s2<400 & s3<400 & s4<400) //
Black,Black,Black,Black BBBB
```

// In this if condition sensors1, s2, s3 and s4 are on black line. There will be no line to track. It will stop moving.

```
{

do

{

digitalWrite(7,LOW);// Relay of motor1

digitalWrite(10,LOW);//PWM1

digitalWrite(8,LOW);// Relay of motor2
```

```
        digitalWrite(11,LOW);//PWM2

    }

    while(s1>400 & s2<400 & s3<400 &s4>400);

    }
```